TECHNOLOGY IN ACTION

TV AND VIDEO TECHNOLOGY

Mark Lambert

The Bookwright Press
New York • 1990

Titles in this series

Aircraft Technology

Car Technology

Spacecraft Technology

TV and Video Technology

Ship Technology

Train Technology

First published in the
United States in 1990 by
The Bookwright Press
387 Park Avenue South
New York NY 10016

First published in 1989
Wayland (Publishers) Ltd
61 Western Road, Hove
East Sussex BN3 1JD, England

©Copyright 1989 Wayland (Publishers) Ltd

Library of Congress Cataloging-in-Publication Data
Lambert, Mark, 1946-
    Television and video technology/Mark Lambert.
       p. cm. — (Technology in action)
    Bibliography: p.
    Includes index.
    Summary: Surveys the electronic principles and the various forms of both video and television and describes the newest developments in the field.
    ISBN 0-531-18327-0
    1. Television — Juvenile literature. 2. Video tape recorders and recording — Juvenile literature.
[1. Television. 2. Video tape recorders and recording.]
I. Title. II. Series.
TK6640.L24 1990
621.388 — dc20   89-35178
                                 CIP
                                 AC

Typeset by Direct Image Photosetting Limited, Hove,
East Sussex, England
Printed in Italy by G. Canale & C.S.p.A., Turin

Front cover A camera crew in operation during the 1984 Los Angeles Olympics.

Contents

1 Instant communications 4
2 Signals through the air 6
3 Radio waves 8
4 Electronic principles 10
5 Sending pictures 12
6 Picture receivers 14
7 Entertainment technology 18
8 Outside broadcasts 20
9 Television and computers 22
10 Magnetic tape 24
11 Video recording 26
12 The video revolution 28
13 From robbery to robots 30
14 Recording sound 32
15 Laser sound and vision 34
16 Audio systems 36
17 Television unlimited 38
18 Multichannel communications 42
Glossary 44
Further reading 45
Index 46

1 Instant communications

The rapid communication of information, news and ideas is vital in the modern world. Two hundred years ago, sending messages over long distances was a very slow process. Information was often completely out of date by the time it arrived at its destination. During the 1800s, the development of both the telegraph and the telephone began to speed up communications. However, because these systems involved sending messages along wires, they could easily break down. A method of sending messages that did not use wires was needed. The development of the wireless, or radio, provided just the answer.

Radio communications became established during the early 1900s. At the same time, people saw that there was a need to send pictures as well as sound. The first workable television systems began to appear in the 1920s. Today, radio and television have many uses, including aircraft and ship navigation, the tracking of aircraft and ships by radar, missile guidance and space technology. But the primary role of radio and television remains that of communicating information and broadcasting entertainment.

Over the years there has been a steady improvement in the technology of radio and television. The principles of our modern

Even this Bedouin tent in the Sinai desert in Egypt is equipped with a television set, providing the occupants with a window on the world not available to earlier generations.

Using the latest equipment, it is now possible to receive a wide range of different television channels. In some households, television has become so important that a special television room is set aside for viewing.

systems remain the same as those worked out in the early 1900s. But the quality of radio and television transmission and reception has been dramatically improved. Today, radio reception is superbly clear and many programs are broadcast in stereophonic sound. Radio and television signals can now be sent and received anywhere in the world. Color television is almost standard throughout the developed world, and with the development of recording techniques, a vast range of recorded audio and video material has become available. At the same time, cable and satellite television are increasing the range of material available to the viewing public. Computer technology is forming an increasingly important part of modern communications systems.

In many parts of the world, therefore, radio and television have become a part of people's everyday lives. We take it for granted that, at the touch of a button, we can listen to or watch something that is happening thousands of miles away. We can even receive amazingly detailed pictures and information from spacecraft located near the farthest planets of our solar system. This miracle of technology has developed from scientific experiments that began over a hundred years ago.

2 Signals through the air

Radio and television work by using radio waves that travel through air or space (see page 8). The existence of such waves was first proved in 1888 by the German scientist Heinrich Hertz. A number of scientists believed that it might be possible to use these waves to carry signals. In 1890, the French scientist Edouard Branly invented a device called a coherer. This invention made it possible to detect radio waves and convert them into electric current. Using this device, the Italian scientist Guglielmo Marconi succeeded in transmitting a signal across the attic room of his parents' house in Bologna, in 1894.

Over the next few years Marconi gradually increased the range of his transmissions. At this stage, however, his radio signals produced just a jumbled noise at the receiver. In 1900 he devised a system that could be used for producing or receiving a signal that could be recognized. Known as a tuned circuit, this device produced a purer signal and increased the range of transmission still further. In 1901 Marconi sent the first signal across the Atlantic Ocean, using Morse code. Since the telephone was already in use, people now wanted to transmit speech by radio waves. In 1906 an American professor, Reginald Fessenden, used a microphone to

Guglielmo Marconi in 1896, shortly after he moved to England from Bologna in Italy. This early equipment generated radio signals by means of a spark generator (on the left in the picture), first invented by Heinrich Hertz.

Above John Logie Baird with the equipment he used to produce the first TV picture in 1926. Using Nipkow disks, he succeeded in transmitting a picture of the head of the dummy "Stukey Bill." Baird later donated this equipment to the Science Museum in London.

Right Technicians in a modern television control room.

make a continuous radio signal vary according to the pattern of sound waves produced by his voice. In 1918 another American, Major Edwin Armstrong, invented the type of radio receiver still used today. This receiver can pick up even fairly weak signals and can be tuned to different stations very easily.

By the 1920s a number of people were starting to investigate the possibility of transmitting pictures as well as sound by radio wave. Two important inventions of the 1880s were the Nipkow disk and the cathode ray tube (see page 12). These inventions eventually led to the development of two very different television systems. In 1926 John Logie Baird, working in Britain, became the first to demonstrate television. But his mechanical system, based on the use of Nipkow disks, was soon shown to have too many disadvantages. This system was replaced by an electronic one, which was developed in America by Vladimir Zworykin.

Radio and television now made progress alongside each other. In 1936, the first live television service opened in London, and in 1953, color television was introduced. The first transistor radio appeared in 1954, followed by the first all-transistor television in 1960. The first stereo broadcasts were made in 1961, and one year later the first television pictures were beamed across the Atlantic by satellite.

3 Radio waves

Radio waves are a kind of energy known as electromagnetic radiation. Other forms of electromagnetic radiation include microwaves, light rays, X-rays and gamma rays. All of these travel at the speed of light and can move equally easily through air or empty space.

Like waves on a pond, radio waves have a wavelength – the distance from the top of one crest to the top of the next. The number of wavelengths passing a particular point each second is known as the frequency. Frequency is measured in cycles per second, or hertz. Because radio waves travel at the speed of light, radio frequencies are very high, and they are generally given in kilohertz (kHz) or megahertz (MHz); 1 kHz = 1,000 Hz; 1MHz = 1,000,000 Hz.

The longest radio waves have wavelengths of around 6,560 ft (2,000 m) and relatively low frequencies of less than 300 kHz. Medium wavelength radio waves have frequencies of up to 3,000 kHz, and the short wave radio band goes up to frequencies of over 20 MHz. Above this there are the VHF (Very High Frequency) and UHF (Ultra High Frequency) bands, which have frequencies of up to 3,000 MHz and wavelengths ranging from 33 ft to 7 in (10 m — 10 cm).

When speech and music are broadcast by radio, the sound signals are transmitted on a carrier wave of a particular frequency. The sound waves are converted by a microphone into a varying electrical sound signal, and this is used to alter, or modulate, the carrier wave. Two kinds of modulated signals are used.

In amplitude modulation (AM) the sound signal alters the height (amplitude) of the carrier wave. AM signals often suffer from interference resulting in hiss and static. This can be almost eliminated by using frequency modulation (FM) in which the frequency of the carrier signal is made to vary. FM signals are used for VHF and UHF transmissions, including television sound

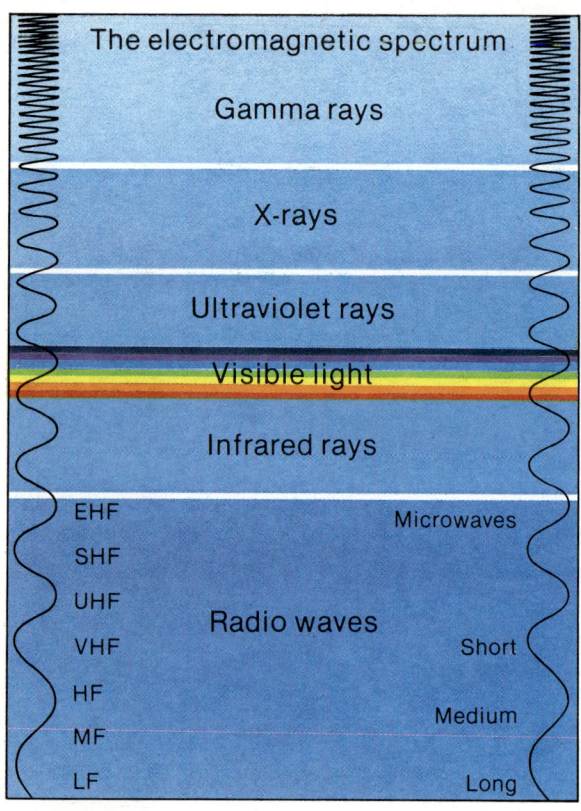

Electromagnetic radiation consists of waves of energy that can travel through space. This radiation comes in a variety of different wavelengths, which together make up what is called the electromagnetic spectrum. Radiation with the shortest wavelengths is known as gamma radiation. X-rays have slightly longer wavelengths and these, in turn, are followed by ultraviolet rays, visible light rays, infrared rays and radio waves. Radio waves can themselves be divided into three bands, short, medium and long waves. All these waves travel at the speed of light. Therefore, as wavelength increases, the frequency decreases. Radio waves are also divided into frequency bands, which range from Extremely High Frequency (EHF) and Super-high Frequency (SHF) to Low Frequency (LF).

signals. AM signals are used for broadcasting long, medium and short wave radio, and for sending television pictures.

Radio waves travel around the world in different ways, depending on their wavelength. Basically all radio waves travel in straight lines, but long waves traveling close to the Earth's surface bend slightly and can be picked up thousands of miles away. Shorter waves can be bounced off a layer in the atmosphere called the ionosphere.

The Earth's surface also reflects radio waves, which can therefore travel around the world by bouncing up and down between the ground (or ocean) and the ionosphere. VHF and UHF transmissions, on the other hand, pass through the ionosphere. They have to be transmitted from one antenna to another.

Right Radio waves can be thought of as being like waves on a pond. The wavelength is the distance between wave crests and each wave crest has a height, or amplitude. The frequency is the number of waves that pass in one second.

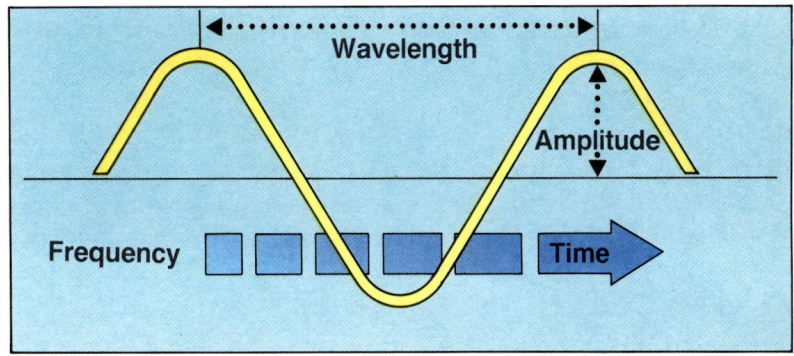

Below Radio waves can sometimes be transmitted around the world by bouncing them off one of the layers of the ionosphere. However, the shortest wavelengths must be transmitted from antenna to antenna or via a satellite.

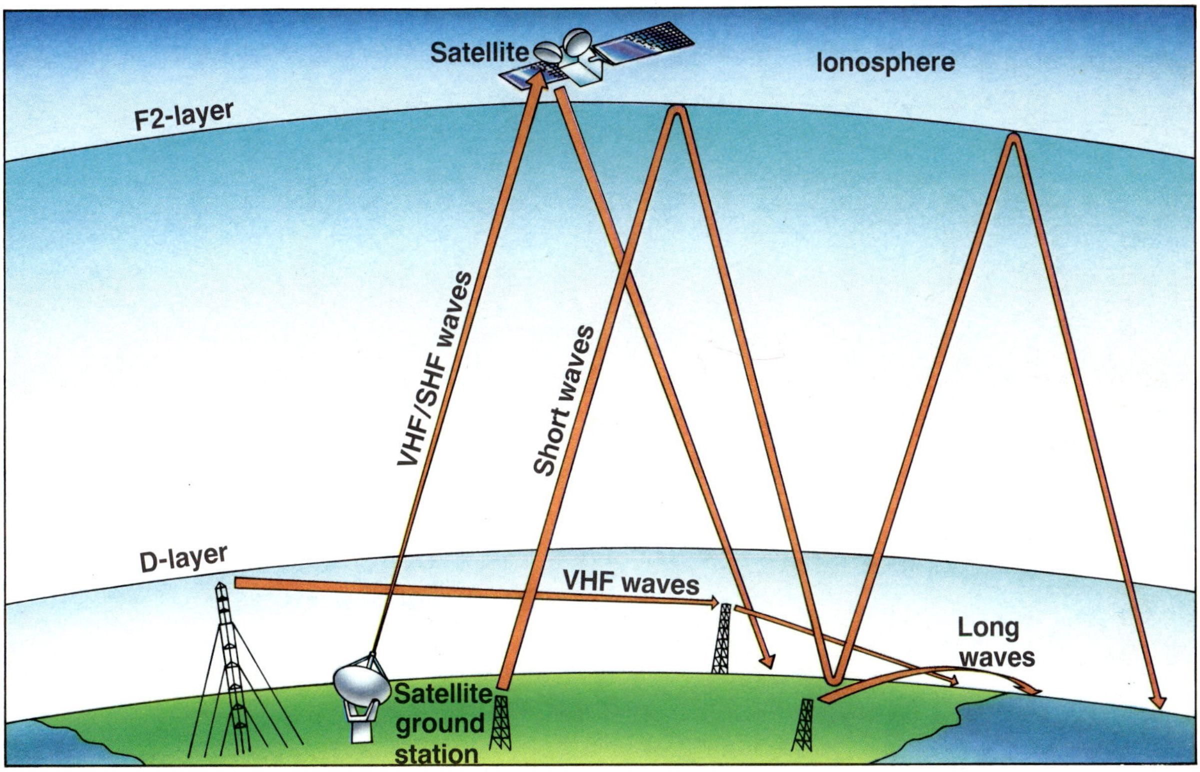

4   Electronic principles

The term "electronic" is very widely used today. Computers, televisions, hi-fi equipment, digital watches, calculators, transistors and silicon chips are all described as electronic devices. But what is it that makes these devices special? Basically, it is because they all work by making careful use of tiny particles known as electrons. These are the particles that, when flowing through a conductor, make up an electric current. The technology known as electronics is the use of devices that control tiny electric currents very accurately.

The very first electronic device was constructed by the American scientist Thomas Edison. In 1883, while experimenting with electric light bulbs, he made what is now known as a thermionic valve. At the time, however, he did not know how it worked and had no idea that it might be useful.

In 1897, the English scientist J. J. Thomson discovered electrons, and a few years later another Englishman, John Fleming, began to see the possibilities of Edison's device. Radio waves produce an electric current that moves to

Guglielmo Marconi equipped his yacht *Elettra* with a large radio laboratory in order to carry out his experiments.

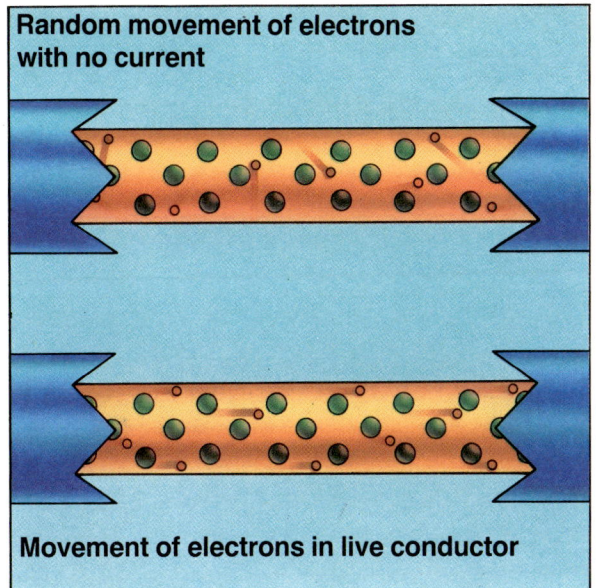

Random movement of electrons with no current

Movement of electrons in live conductor

In a metal conductor "free" electrons (red) move randomly between the atoms of metal (green). If the conductor is connected to an electrical supply, these electrons tend to move in one direction only, and an electrical current flows along the conductor.

and fro. This current, called an alternating current, is difficult to detect. Fleming realized that Edison's device could be used to change alternating current into a direct current, which moves in one direction only and can easily be detected. In 1904, Fleming constructed a new device, known as a diode valve, which he used successfully in a radio receiver.

In 1906, an American scientist named Lee de Forest devised another type of valve, known as the triode valve. He used this to make the small electric current produced by a weak radio signal vary a much larger current produced by a battery or mains supply. Thus the triode valve could be used to increase the power of, or amplify, a radio signal. Radio receivers now became much more sensitive.

The triode valve could also be used as an electronic switch – using one electrical signal to switch another signal on or off – and such valves were used in early computers. However, they were unreliable and used a great deal of power. In 1947 William Shockley and a team of scientists in the United States invented the transistor, making use of silicon. This device could perform the same tasks as the triode valve and soon began to replace it.

Transistors have also replaced the valves formerly used in televisions. The all-important picture tubes of televisions were developed from yet another electronic device, the cathode ray tube, which was invented in 1897 by the German scientist Ferdinand Braun.

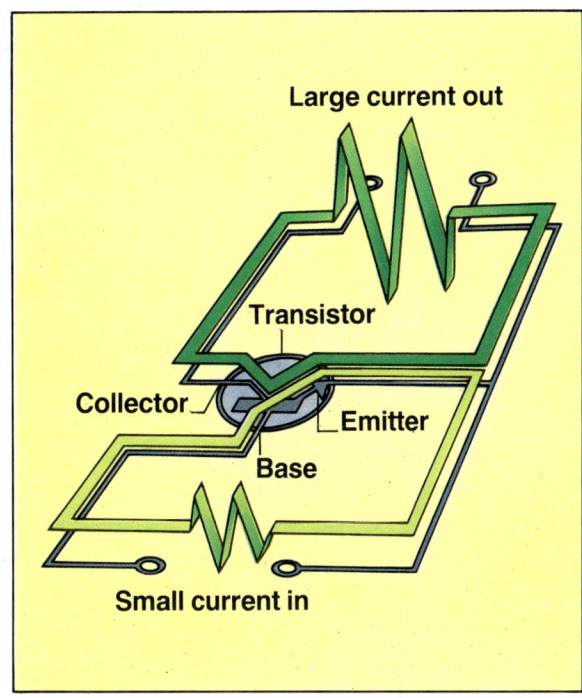

A transistor can be used to amplify, or increase the strength of, an electrical signal. Inside are three layers of material. Two of these, the collector and the emitter are separated by the third, the base. The amount of current that can flow between the collector and the emitter depends on the amount of current that is fed to the base. The signal to be amplified is fed into the base terminal of the transistor. The variations in this signal cause much larger variations in a current flowing between the collector and emitter terminals of the transistor.

5 Sending pictures

The basic technique used to send a television picture is very simple. First, a system of lenses, like those used in any camera, is used to create an image of the subject. This image is then delivered to a camera tube, where it is converted into electrical signals that can be recorded or broadcast.

The tube of a television camera is based on the cathode ray tube. The image created by the lens system is focused onto a special screen at one end of the tube, where it is "read," or scanned by an electron beam. The screen is made of a material that conducts electricity in amounts that vary according to the brightness of the light. It consists of thousands of tiny picture cells, or pixels. Each pixel provides

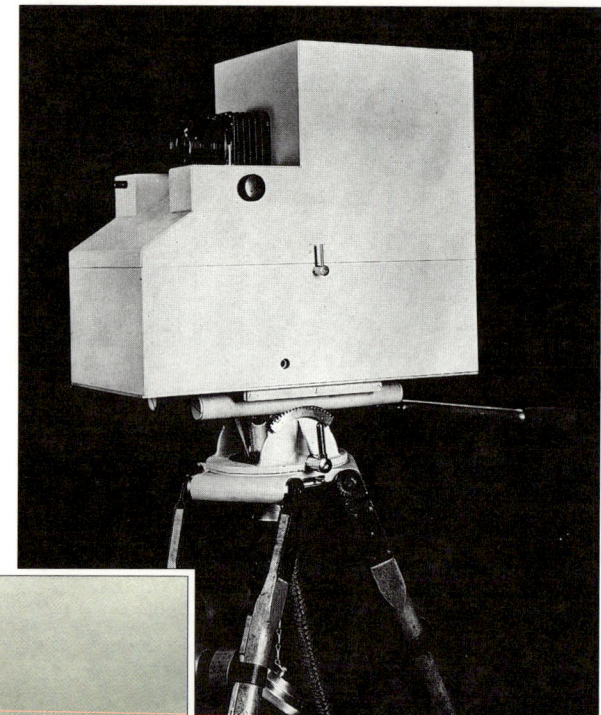

Above One of the Marconi-EMI Emitron television cameras used to launch the world's first public high definition television service in November 1936. Such cameras had very limited movement.

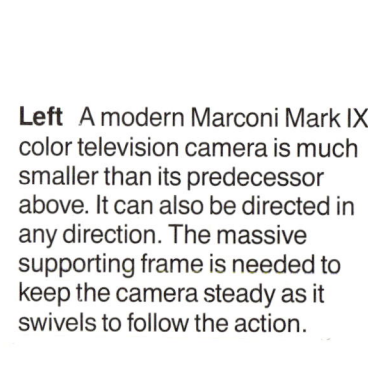

Left A modern Marconi Mark IX color television camera is much smaller than its predecessor above. It can also be directed in any direction. The massive supporting frame is needed to keep the camera steady as it swivels to follow the action.

12

information about the brightness of a tiny part of the image. The pixels of a modern television camera are usually arranged in 625 or 525 lines. An electron beam generated at the other end of the tube scans the image line by line 50 times each second. As the scanning beam passes along the lines of pixels, it generates an electrical signal that varies according to the pattern of light on the screen.

The earliest type of camera, developed by Vladimir Zworykin, used a type of tube called an iconoscope. This tube was later replaced by the orthicon, which produced a more detailed picture signal, and then by the image orthicon, which was very much more sensitive to light. Modern television cameras use a type of tube known as a vidicon. One of the main advantages of this type of tube is that it can be made very small, which makes it possible to use smaller, cheaper lenses in the camera.

A black-and-white television camera uses just one camera tube. In a color camera there are usually three or four tubes. Before the image created by the lens system reaches the camera tubes, it is split into its three primary colors (red, green and blue) by a system of special mirrors. Each camera tube generates a signal for a different color.

In order to create the signal to be recorded or broadcast, the three signals from the camera tubes are passed to a pair of electronic devices. One of these creates a black and white light intensity signal, the other creates a color quality signal, and together they combine the three signals into a single video signal. A timing signal is then added, and the complete signal is passed either to a video recorder in the studio or to a transmitter. The sound signal, which is generated separately, is recorded or broadcast at the same time.

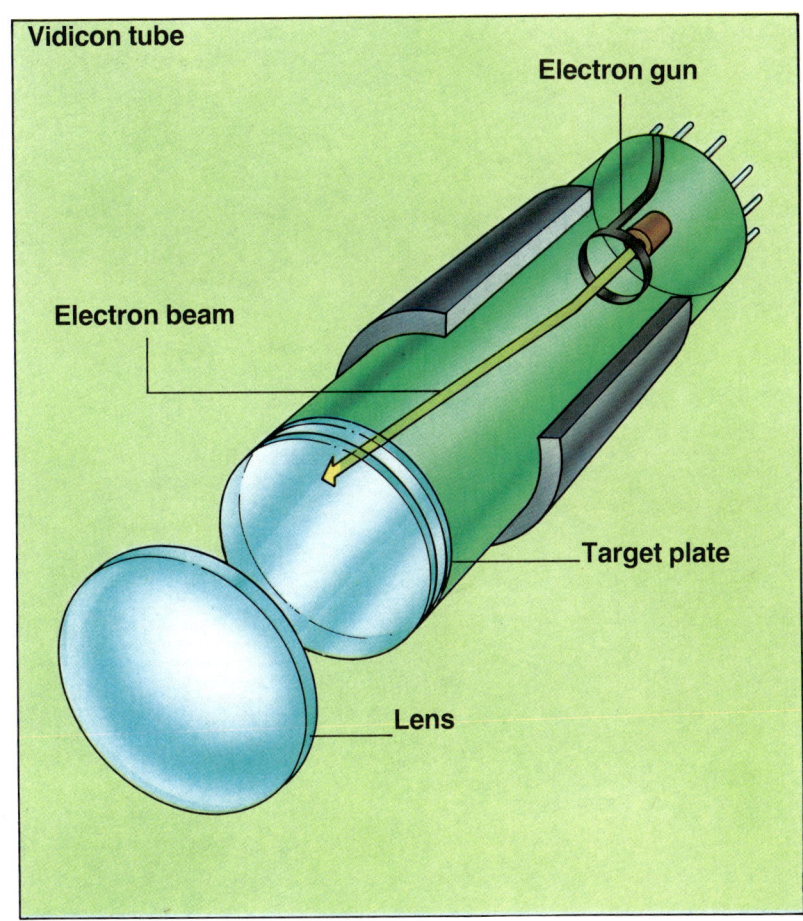

The vidicon tube is the basis of all today's television cameras. The lens forms a visible image on the light-sensitive target plate, which is then scanned by an electron beam. Each pixel (picture cell) on the target plate generates a tiny electrical current, the strength of which depends on the amount of light in the image. Color information can be obtained by using three vidicon tubes together with color filters. However, in cameras where space is limited, such as camcorders (see page 28), a single tube is used. The filtering process occurs just in front of the target.

6 Picture receivers

A television receiver reverses the process that occurs in a television camera. Here, the broadcast video signal is changed back into a continuously moving image on the screen. The screen is at one end of the picture tube, which, like a camera tube, is a form of cathode ray tube.

The broadcast signal is picked up by the television antenna. From there, an antenna cable carries it to the receiver. The antenna cable has a central copper wire and a tube of copper mesh separated by a thick layer of insulation. The signal is carried by the central wire, and the tube shields the signal from outside electrical interference.

At the receiver, electronic circuits divide the signal into a timing signal, a light intensity signal and a picture signal. In a black-and-white television there is just one picture signal, which corresponds to the signal generated in the camera. The signal is fed to an electron gun, where it causes changes in the strength of an electron beam, which scans to and fro across a fluorescent screen at the far end of the picture tube. The changes in the beam's strength produce lighter and darker regions along each

The television pictures received in your home can arrive in several different ways. The signals may come direct from the broadcasting company, often being boosted, or amplified, on the way by relay stations. Some stations transmit their signals to satellites, which then rebroadcast them down to antennas on the ground. In some places television pictures come via a cable from a cable television company.

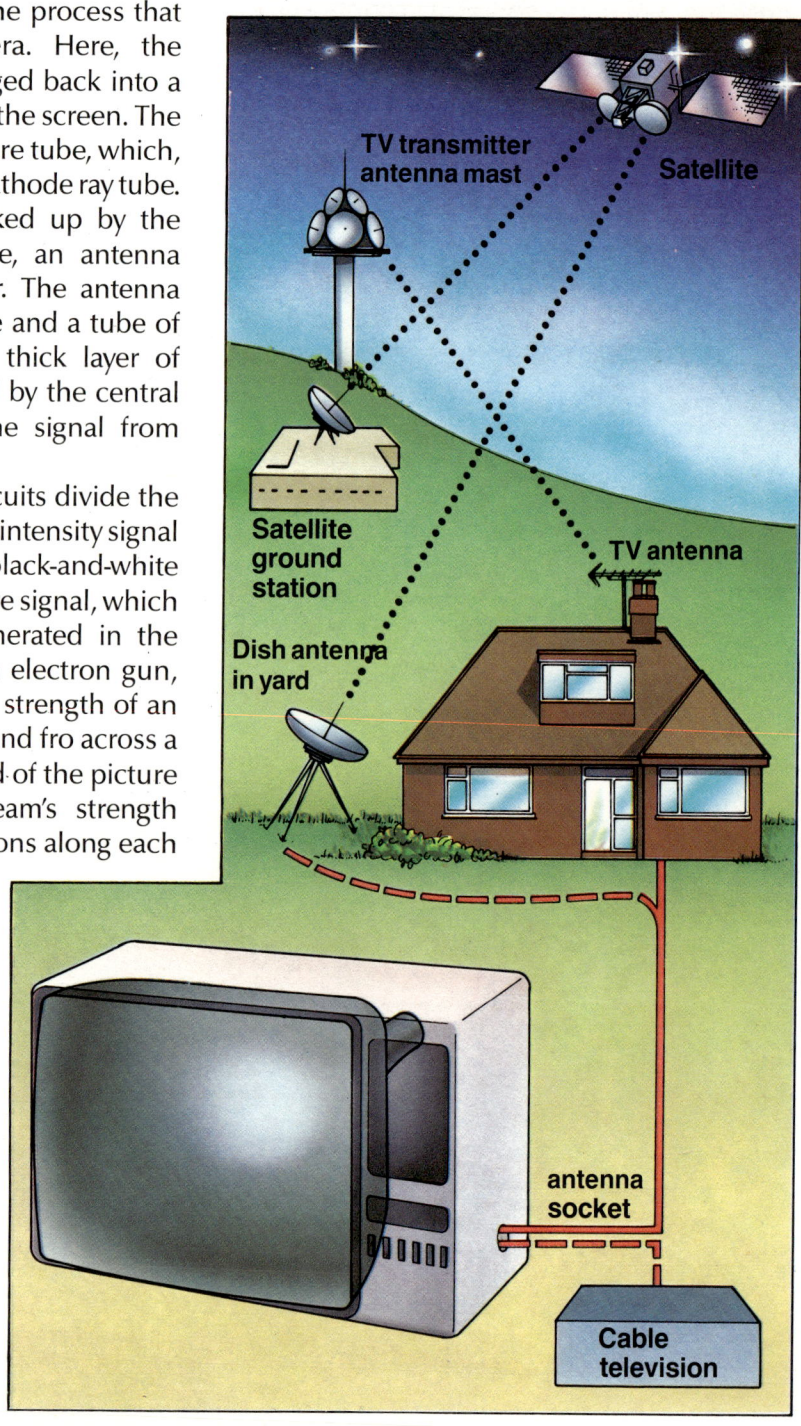

14

line traced by the beam, and a picture is built up over the whole screen. Each spot on the screen is scanned twenty-five times a second. Because this happens so quickly, the eye sees a constantly moving picture.

A color television works in a similar way. The timing signal is first separated from the picture signal, which is then used to generate three separate color signals – red, blue and green – and a light intensity signal. The separated signals are fed either to three separate electron guns or to a single gun that generates three beams. These are fired at the screen at the same time, scanning it 50 times each second.

The color picture is created by making the beams pass through a grid, known as a shadow mask, before they reach the screen. The screen has an arrangement of three kinds of phosphor dots or lines, which glow red, green or blue when triggered by the arrival of electrons. Because of the way in which the holes or slots in the grid are arranged, each beam triggers only one kind of phosphor. The phosphors glow more or less brightly according to the strength of the signal they receive. These variations in strength produce constant changes in color.

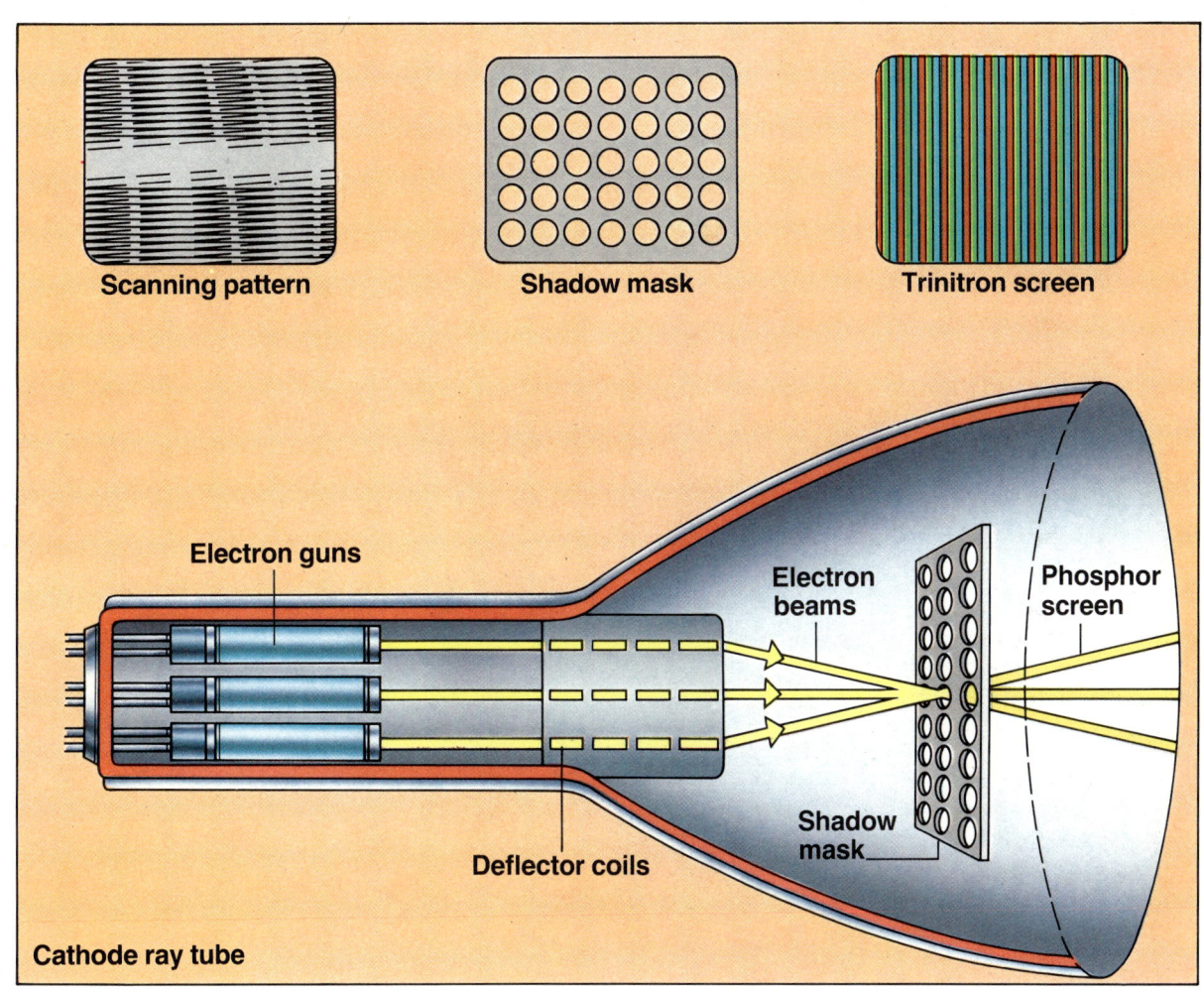

In a modern color television set three electron guns generate beams that scan the screen in a regular pattern. A shadow mask ensures that each beam strikes the phosphor lines of only one color.

Modern color televisions come in a wide range of different sizes and types. Screen sizes, which are always measured across the diagonal, range from 35 in (89 cm) down to just 2 in (5 cm). Small screen sizes of a few inches have been made possible by the increasing miniaturization of electronic components. However, the smallest screens work in an entirely different way, using LCD (liquid crystal display) screens instead of cathode ray tubes.

A liquid crystal is a material that changes its shape when it receives an electrical voltage. As its shape changes, the way in which it transmits light also changes. Such crystals are used in calculator and digital watch displays. The screen of an LCD television is made up of thousands of tiny picture cells, each of which contains a liquid crystal. Just as each phosphor dot of a tube television receives a signal that makes it glow, each picture cell receives a signal that makes it transmit more or less light from a backlight behind the screen. Color filters on the picture cells produce a color image. At present it is possible to build only small liquid crystal televisions, but LCD technology is improving all the time.

In recent years the quality of television pictures has been greatly improved. This is due partly to improvements in transmission and partly to improvements in television sets themselves. Among such improvements are black screens (instead of gray ones), better phosphors, better electron beams and better beam focusing. FST (flatter, squarer tube)

A miniature LCD television combined with an 8mm video recorder into a single unit. It can be used to record transmitted programs or play back video tapes that have been recorded, using an 8mm camcorder.

16

PICTURE RECEIVERS

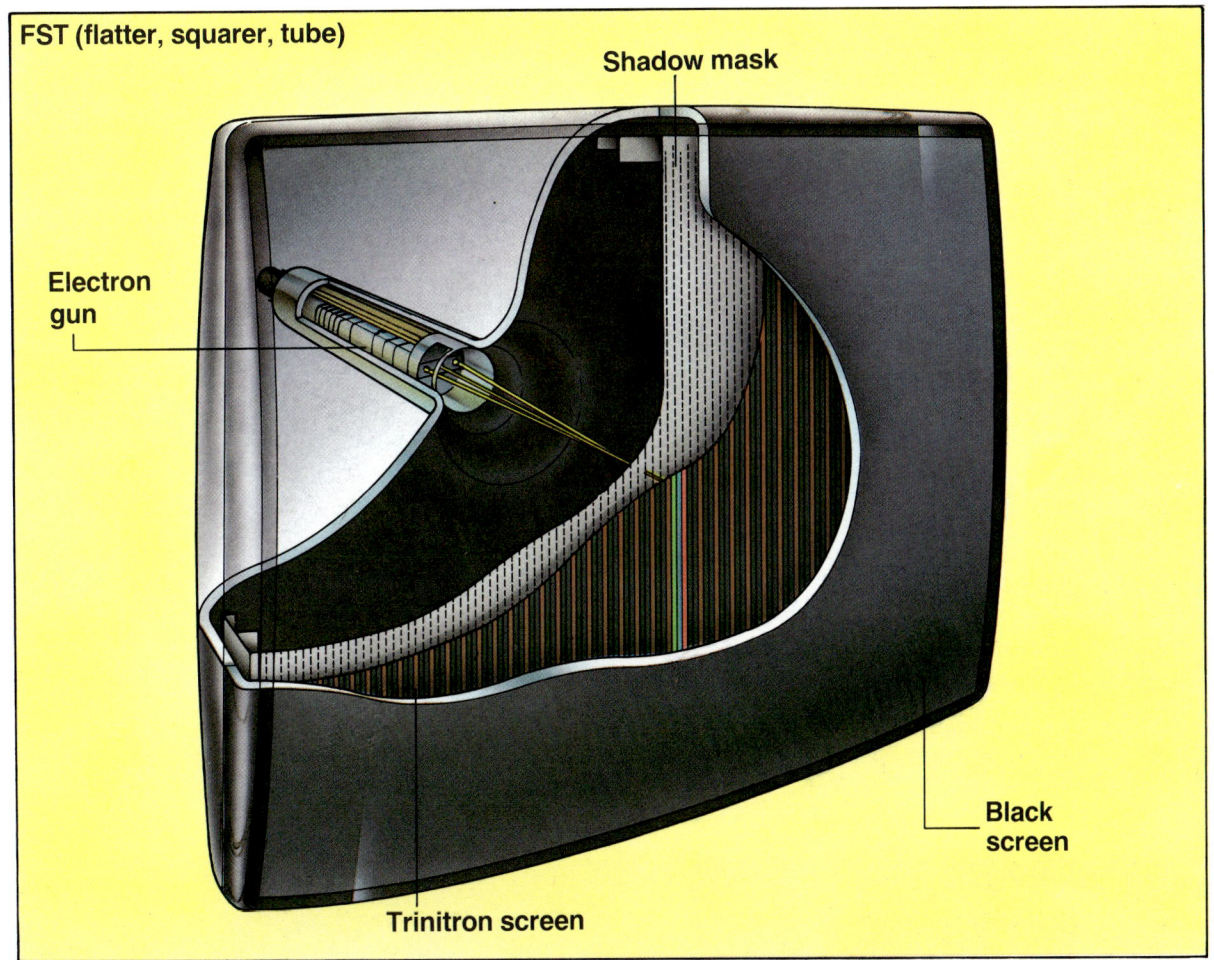

In the most recent color television sets a single electron gun generates all three electron beams. These are focused onto the phosphor lines on the screen, passing through the shadow mask on the way. The black coating on the screen combined with black tinted glass helps to improve the contrast and reduce reflections.

televisions produce less picture distortion than the older more rounded screens, and make it possible for people to view the screen more easily from an angle.

Many television sets now come equipped with a remote control, which allows the viewer to operate the television from the comfort of an armchair. Each push button on this control generates a beam of infrared signals that are picked up by a sensor on the set. One of the latest features of television is known as picture-in-picture. This enables the viewer to watch one channel and at the same time see what is taking place on another channel on a small picture, inset into the screen. A choice of four positions on the screen means that the inset need not obscure essential parts of the main picture. Some televisions are already using microchip technology to allow the viewer to freeze the action on the screen via the remote control. In the future it will be possible to zoom in on a particular part of the picture or obtain an instant action replay of any chosen sequence. Television technology is changing very rapidly.

7 Entertainment technology

Most television is concerned with making and broadcasting programs as entertainment for the viewing public. Some programs are recorded on video tape (see page 24) for broadcasting at another time; others are broadcast live.

Many programs are made in television studios. A television studio is basically just a large, tall room, with doors large enough to enable almost any item to be moved in and out easily. In one or more areas are the sets, where the action to be filmed takes place. The layout and type of set depend on the type of program being broadcast, which can range from a news and current affairs program to an educational program, quiz show or drama. The action is recorded by a number of cameras, which glide silently and easily around the studio floor. The camera operators receive instructions via headphones from the director who sits in a control room overlooking the studio. The director also gives instructions to the studio floor manager, who organizes everything that goes on in the studio. Television presenters often read their lines from teleprompt screens mounted on the cameras. They may also wear small earphones, so that they can receive messages from the director.

Making a television program involves a large number of people. In addition to the actors, there are the director, the cameraman, sound engineers and lighting technicians.

Other studio personnel include the people whose job it is to move scenery and props. Electricians operate the studio lights, many of which hang from the ceiling. Sound engineers control the microphones used to pick up sound. Sometimes overhead microphones are used, particularly for drama. In other types of programs, radio microphones are often used. These do not require long, trailing cables and are therefore very useful when people have to move around a great deal.

Upstairs in the gallery, the director sits with other controllers in front of a control panel, or mixing desk. In front of the control panel are several monitors, or screens. Some of these show the pictures being produced by the studio cameras; others show such things as timing clocks, captions and material that is held on film or videotape.

The picture that is to be broadcast is created by a person known as the technical director, who switches from one monitor picture to another on instructions from the director. A production assistant controls the timing of the various items being broadcast, and engineers control the quality of the pictures and sound.

In the control room, the director watches what is going on in the studio by means of an array of television monitors.

A live broadcast involves a large number of people who remain unseen by the television viewer. Cameramen and sound engineers operate the equipment, while the floor manager makes sure that everything runs smoothly.

8 Outside broadcasts

A number of occasions and events that appear on television take place far away from studios. Such things as sports events and concerts have to be televised where they happen.

One way of doing this is to record the event for later broadcasting. This used to be done using film, which could be loaded onto a special projector for broadcasting. However, before it can be shown, film has to be chemically processed and then edited by physically cutting it up and sticking the pieces together in the required order. This complicates and slows down the process of showing a filmed report.

In recent years, however, the film camera has been replaced for many purposes by its modern equivalent, the video camera. This is a small portable television camera that records pictures onto magnetic tape (see page 24). The tape requires no chemical processing and can be edited much more rapidly by electronic means. A video news report can therefore be shown within minutes of its arrival at the news studio.

Today, all outside broadcasts can be viewed live on television. At this car racing event, pictures from each camera are carried by a cable to an outside broadcast van, from which they are transmitted to the television station.

In 1988, President Reagan and Premier Gorbachev held a summit meeting in Moscow. The event was of such importance that television pictures were transmitted all over the world.

Alternatively, events can be broadcast live from where they are happening. This is not a new idea; in 1937 live pictures of the coronation of King George VI of England were broadcast and in 1939 live pictures of the opening of the World's Fair in New York showed the arrival of President Franklin D. Roosevelt. Since then live outside broadcasts of events have become more and more common, and audiences now take them for granted.

Early outside broadcasts involved the use of heavy, stationary cameras, many miles of cable, and tons of expensive equipment to provide the power for transmission. As the demand for outside broadcasts continued, the cameras, generators and transmitters were made lighter and more efficient. Today's equipment is a great deal more mobile and easier to use. Cameras are now small enough to be carried on the shoulder and can even be operated by someone riding on the pillion seat of a motorcycle.

The modern portable camera requires no cables; the pictures may be recorded on videotape and then quickly taken back to an ENG (electronic news gathering) van. Here, using power produced by diesel-driven generators, the signal is transmitted to the television station for live broadcasting on the television network. Sometimes a camera is equipped with a small portable antenna. This transmits live pictures to the ENG van, where they are amplified and retransmitted to the television station.

9 Television and computers

Television and computer technologies originally developed alongside each other. Early computers processed information stored in the form of punched cards or tape, and a printer produced the results in a typewritten form. However, by the late 1940s magnetic tape was being used to store computer data, and scientists were starting to develop the cathode ray tube as a screen. Now the information stored in a computer memory could be simultaneously viewed and altered.

Today's computers use screens, known as monitors or VDUs (visual display units) that work exactly like television screens, and an ordinary television set can often be used as a computer monitor. The difference is that when acting as a computer monitor, the television tube receives signals directly from a computer instead of broadcast signals via an antenna. Many computer monitors are designed only for

Computer monitors are television screens that receive signals from computers.

their one particular task. The information displayed on their monochrome (single color) screens often appears green or orange. The color depends upon the fluorescent material used to make the screen.

Computers do not have to be linked directly to a monitor. Videotex systems are computer systems which enable information stored on large computers to be displayed on TV screens or computer monitors in people's homes and places of work. Two types of videotex systems exist. One of these is known as teletext, which is used to display "pages" of information on television. The information is compiled by television companies and broadcast along with ordinary television programs. The signal occupies a few spare lines at the top of the television picture, normally just off the screen. If the television set is equipped with a decoder, the viewer can display any "page" of the broadcast information simply by pressing buttons on a key-pad. Some modern television sets use a system known as Fastext. Microchip memories make it possible to store and call up teletext pages more quickly.

A "page" of Teletext information.

The second type of videotex system is known as viewdata. Information, supplied by such organizations as banks, airlines, newspapers and major retail shops, is stored on a central computer, by the operators of the viewdata service. Anyone wishing to use the information must first register with the viewdata service and pay a fee. Then, using either a television screen and a key-pad, or more usually, a micro-computer screen and keyboard, the viewer can call up the information, which is sent in coded form via the telephone system.

An ordinary color television set can be tuned into a computer and used as a monitor. Here, a television set is receiving signals created by an electronic graphics pad linked to the computer.

23

10 Magnetic tape

A flowing electric current generates a magnetic field, and this fact is used by all today's magnetic recording devices. A sound signal is made to produce a varying electrical signal, which is then recorded as a varying magnetic pattern on continuous plastic tape coated with tiny crystals of iron oxide.

In a tape recorder, the magnetic tape is wound at a constant speed past two electromagnetic heads. When the machine is recording, the tape first passes over an erase head, which removes any previous recording. The record/play head then stores sound signals on the clean tape by arranging the tiny crystals into a pattern. During playback, the record/play head is used to "read" the magnetic pattern and turn it back into sound signals.

Magnetic tape is also used for recording television video signals. However, there is more information in a video signal. The tape is therefore wider, and material is recorded in tracks that run diagonally across the tape (see page 27).

Recording sound in this way is known as analogue recording. This is because the recorded pattern corresponds to (is analagous to) the original sound. However, such recordings are often distorted, as unwanted signals become included in the main signal during recording or playback. To overcome this,

This DAT (digital audio tape) machine is one of today's most compact high quality recording systems, one of the bulkiest items being the microphone needed to generate the required electrical signal.

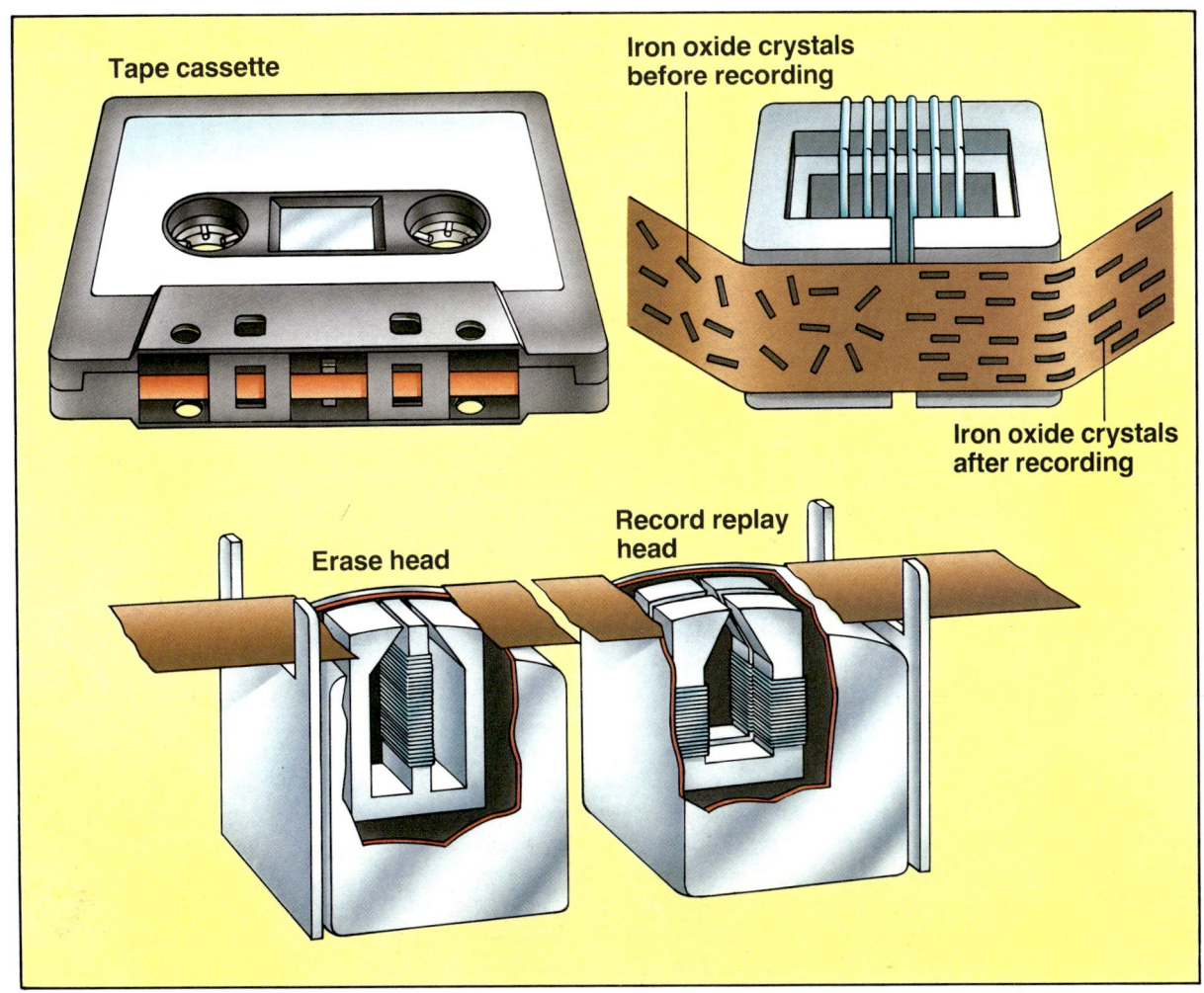

In a tape recorder the record/replay head generates a varying magnetic field that rearranges the crystals of iron oxide on the tape into a pattern. During replay, the head converts the pattern back into an electrical signal.

many manufacturers are turning to digital recording, in which the signal is converted into a series of pulses that can be represented by numbers, or digits. The digital representation of the sound is made by sampling the varying analogue signal at regular intervals. Each sample is then given a digital value and converted into binary code, the two-digit language used by computers. The main advantage of this system is that the signal can be analyzed by microchips and altered in such a way as to improve the signal that is eventually played back.

Digital recording is used in producing laser audio and video disks (see page 34). It is also being used to record on a new form of magnetic audio tape known as digital audio tape (DAT). This is a high quality tape, similar to that used for video recording. Rotating heads record tracks across rather than along the tape, and a large amount of digital information can be recorded very accurately, resulting in near perfect recording. At present, however, DAT equipment is very expensive, and recording companies are concerned about the possible use of DAT to make illegal copies of recorded material.

11 Video recording

Video recorders first appeared during the 1960s. However, they did not sell in large numbers until the 1970s, when several companies developed video cassette recorders. These used removable tapes that were safely contained in plastic cassettes designed for the purpose.

The designs of the first video cassette recorders differed greatly, and it was impossible to play one manufacturer's tape on another manufacturer's machine. Philips produced the VCR (Video Cassette Recorder) system and although this later disappeared, the term VCR is still used to describe all video recorders. For a while the Betamax system produced by Sony dominated the market, but today VHS (Video Home System), originally developed by the Japanese company JVC, has become established as the most widely used system.

VCRs have freed television from one of its main drawbacks. People are no longer limited to watching television programs as they are broadcast. Viewers can now record programs for viewing at a later time. In addition, they can watch any other prerecorded material of their own choosing. Like many television sets, most modern video recorders are controlled via infrared remote controls. At the same time, video recorders have for some time had a number of useful features. Most video recorders allow the user to search through the

Bar codes for use by video recorders can be created using a computer.

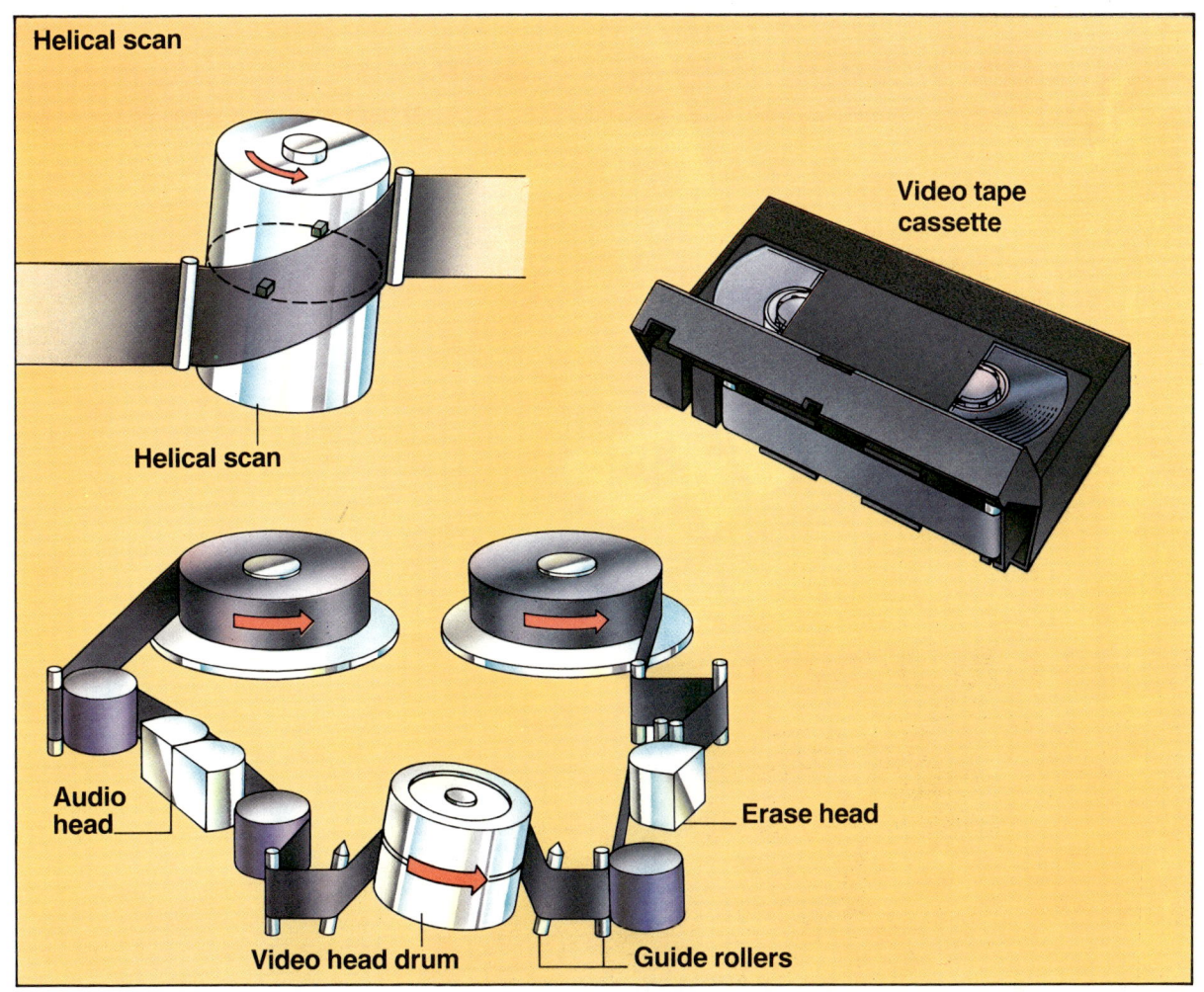

The high quality tape used in a video recorder is carefully protected inside its cassette. In the recorder a loop of tape is lifted out and passed over a pair of tilted recording heads that record tracks at an angle across the tape.

recorded material, using fast-forward and rewind controls. Some have slow motion and freeze frame facilities, which can be used to slow down the action or stop it at a particular moment. On some video recorders the freeze frame facility is just a way of pausing the tape. Others, with more sophisticated electronics, give a perfect still picture.

Another feature of video recorders is their ability to record programs while their owners are out for the evening or away on vacation. Programmable timers make it possible to set the machine to make recordings two or sometimes four weeks ahead. Recording times are normally programmed by pressing buttons to set the start and finish times on a digital display. On some machines this can be done using the handset, and there is at least one machine with a handset that gives a "spoken" confirmation of the recording times set. Some of the latest kinds of VCRs use a laser to "read" bar codes, which set the times automatically. At present the bar codes are printed on a special card, but there are plans to include the codes in the magazines that publish the times of television programs.

12 The video revolution

The early video recorders of the mid-1970s were expensive, with limited facilities and poor picture quality. At the same time the range of material available for viewing was small. Today, however, as a result of a steady improvement in the electronics of these machines, picture quality is excellent and most video machines have a number of useful features. Stores and video libraries stock a wide range of films, and many of these are made especially for the video market. As a result, an ever increasing number of television owners now own a video recorder.

Many viewers use their video recorders largely for recording broadcast television programs for viewing at another time. On the other hand, prerecorded video tapes have also proved to be very popular. At one time it was thought that the sale of taped films might greatly reduce the size of movie theater audiences. But in fact today's movie theaters continue to attract audiences; many people still prefer to see films on a large screen.

Video cameras have now superseded ciné cameras for recording home movies. Unlike ciné film, magnetic tape does not have to be developed and can be replayed instantly. In addition, magnetic tape can record sound as well as pictures; sound recording with ciné film was always expensive and difficult to achieve. As a result video cameras are increasingly being used to make a permanent record of events such as weddings.

Video systems have a wide variety of uses. In this Tokyo beauty salon the beautician uses the video screen to superimpose different make-up styles or hairstyles on the image of a client, so that the client can choose which style she prefers.

Video technology has also made the instant home movie possible. Ciné film cameras are rapidly being replaced by small portable video cameras, or camcorders. These can record images that can be viewed immediately, without the need for developing the film.

A modern camcorder is a combination of a miniature television camera and a video recorder. As in any other television camera an electrical signal is created by using a lens system and a tube or chip. This signal is passed directly to the recording head, where it is transferred to magnetic tape. An audio signal, produced by a microphone, is recorded at the same time.

As with the early VCRs, different manufacturers have brought out camcorders that operate using very different designs. The smallest and lightest camcorders use 8 mm tape in cassettes only slightly larger than audio cassettes. These 8 mm tapes cannot be played back using standard VHS video machines, although they can be displayed on the television set directly from the camcorder. VHS-C camcorders, on the other hand, use standard VHS tapes in small cassettes, which can be played back using a special adaptor in a VHS video recorder. Recently, technological advances have made it possible to produce VHS camcorders that use full-size VHS tapes.

Camcorders are rapidly becoming an important part of the photographic market. Manufacturers are now offering equipment for transferring old home movies on ciné film, and even still transparencies, to video tape.

13 From robbery to robots

Outside the home, television and video have a number of other uses. These range from helping to prevent shoplifting to providing robots with "sight."

Theft and robbery from stores, banks and other premises have unfortunately become all too common in recent years. To counter this threat, many such places have installed closed circuit television, so called because the pictures are not broadcast. Instead, they are displayed on screens that form part of the immediate system. Remote-controlled cameras record everything that happens in the store or bank, and the pictures are displayed on screens placed where they can be seen by the staff or by security guards. In many cases a video recorder fitted with a continuous loop of tape makes a recording. In this way there is always a record of the events of, for example, the previous half-hour.

Closed circuit television is also used by the police to monitor traffic on busy roads and large gatherings, such as those at sports events. Nuclear power workers can safely monitor dangerous materials, such as stored radioactive waste. Video conferencing combines closed circuit television with the telephone system to allow business people in different places to confer "face-to-face" with one another.

Closed circuit television is useful for surveillance. Here, a security man watches a bank of screens that monitor what is happening on the platforms of a London Underground (subway) station.

One of the ways in which a robot could be made more useful is to provide it with a television camera. However, teaching a robot's computer "brain" to recognize different objects, such as chess pieces, is very difficult.

Television and video have also proved to be extremely useful aids to learning. Television companies broadcast educational programs for both schoolchildren and adults. And a great deal of teaching material is also available on prerecorded tapes. Businesses often use video tapes to help train new staff. Athletes and dancers use video to help improve their performance.

Television is also used for military purposes. For example, some air-to-air missiles home in on their targets using a remote control television camera. The pilot of the attacking aircraft locks the missile onto the target using a television monitor in the cockpit. Television has also played an important part in space exploration, using remote control machines such as the Russian Lunokhod roving vehicles that explored the Moon and the American Viking landers on Mars.

Today, engineers are working on ways of making robots that can see. One way of doing this is to provide a robot with a television camera and a device for converting the information from the camera into a pattern that can be stored by the robot's computer "brain." This is the easy part; robots can be programmed to recognize simple objects under particular lighting conditions. But scientists are still working on ways of making a computer understand all the information that can be recorded by a television camera.

14 Recording sound

Today's sound recording techniques have evolved over a period of more than 120 years. The first recordings available to the public were made on cylinders wrapped in tinfoil, a system, developed by the American inventor Thomas Edison in 1878. In 1886 the wax cylinder was invented, and in 1888 Emile Berliner invented the first flat disk. Other milestones in the development of sound recording include the invention of the microphone (1925), the development of the long playing (LP) record (1948) and the introduction of the stereophonic record (1958).

Modern recordings are made in specially designed studios. Here the sound is captured using a number of microphones, and the results are recorded on a multitrack tape. Engineers then edit and adjust each of the tracks, finally mixing them to produce the finished master tape. The master tape can then be copied onto cassette tapes or used to make compact disks (see page 34) or records.

The first stage in the production of a record is the making of a master disk. The sound recording on the tape is played through a special lathe. The lathe cuts a spiral groove on a lacquer coated aluminum disk rotating at 33⅓ or 45 r.p.m. (revolutions per minute). The groove is very small, and successive spirals are very close together – up to 140 per 0.4 in (1 cm). The cutting tool is V-shaped and is generally made from sapphire or diamond. Most modern records are stereo recordings, in which the sound signal is divided into separate left and right channels. We normally use both ears to hear sound, so stereo recordings are more realistic. The two signals are fed to the cutting tool, which records them on opposite sides of the groove.

The lacquer master disk is then coated with nickel, using a process known as electroplating. The nickel coating is peeled off, revealing a negative of the master disk on its surface. This negative is again coated with nickel to produce a

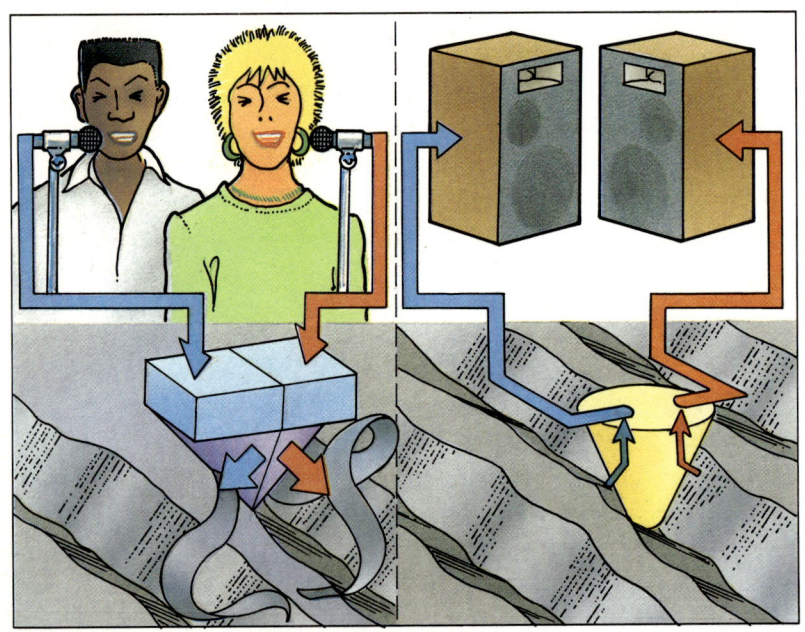

Stereophonic sound is recorded by dividing up the original sound into two signals and using them to record two separate patterns. When cutting a record, the cutting stylus is made to vibrate in such a way that the signals are recorded on opposite sides of the groove. When the record is played back, the two signals are picked up by the stylus and sent via the amplifier to two speakers. The listener hears the sound coming from two different places, which is what would happen if the sound were "live."

32

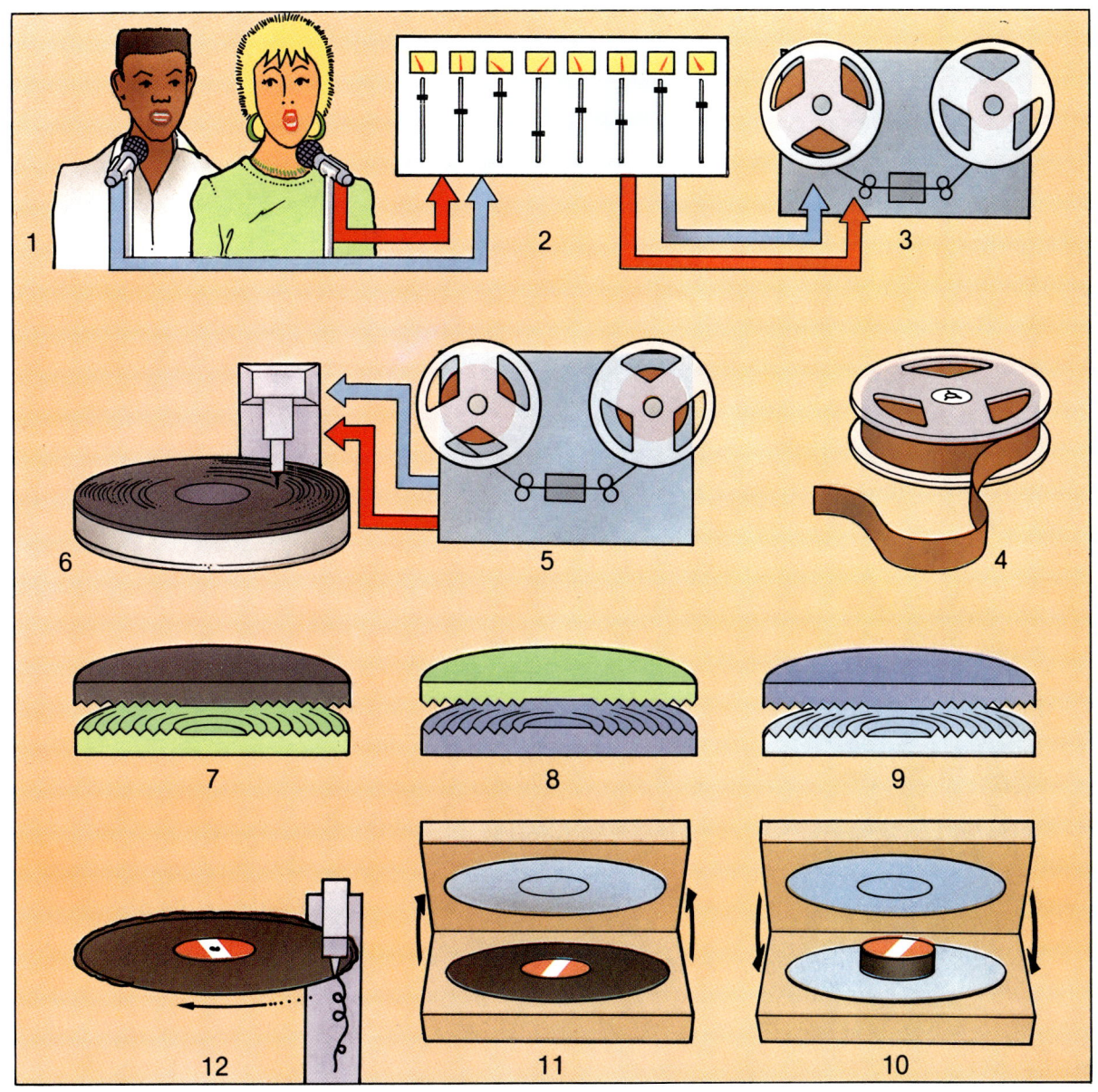

Making a record: 1 Recording, 2 Mixing the sound, 3 Multitrack tape, 4 and 5 Stereo master tape, 6 Disk cutting machine cuts groove onto a lacquer coated blank disk, 7 Nickel negative, 8 Nickel positive, 9 Negative stamper, 10 and 11 Stamping the record from hot PVC, 12 Trimming the finished record.

positive copy of the master, which in turn is used to make a negative metal stamper. The final record is made by pressing a piece of hot, soft plastic between two metal stampers, one for each side of the record.

The record is played back by causing a stylus, or needle, to travel along the groove. As it does so, the recorded pattern on each side of the groove produces vibrations in the stylus. The vibrations in the stylus are converted into electrical signals, which are then converted into sound by the record player.

15   Laser sound and vision

There are two main drawbacks to using conventional records. First, it is impossible to achieve a perfect reproduction of the original sound. Dust, scratches, vibrations from the surroundings and imperfections in the turntable and pick-up arm all contribute to poor reproduction. Second, on a record that is played frequently, the constant rubbing of the stylus along the groove tends to wear both of them out and reproduction becomes poor.

To overcome these problems, sound engineers developed a system that uses a totally different kind of disk, known as a compact disk. On this type of disk the audio signal is recorded in the form of millions of tiny pits in a shiny, reflective surface. The disk, supported on a cushion of air, spins at high speed. The pits are scanned by a beam of laser light, and the reflected beam is converted into an electrical signal. The pits are protected by a tough layer of plastic, and as nothing mechanical ever touches them, they are not worn away. At the same time the laser beam is focused only onto the pits and remains unaffected by small amounts of dust

A laser disk. The disk appears shiny because it has been coated with a highly reflective layer. In the laser disk player this reflects the laser beam back from the surface, and thus generates the signal pattern recorded in the form of pits. The pits are so tiny and so close together that a phenomenon known as diffraction of light occurs. Light rays are bent slightly, and white light is split up into its component colors. This is why the surface reflection includes a rainbow effect.

or minor scratches on the surface of the plastic. Any errors that do occur are corrected by a small computer in the compact disk player. Error correction is so good that a hole 0.08 in (2 mm) wide in the surface of the disk may go unnoticed by the listener.

As with an ordinary record, a compact disk is made in a series of stages. First, the analogue audio signal is converted into a digital signal. Next a master disk is created by using the signal to make a laser beam "cut" a spiral of pits into a smooth layer of light sensitive chemical on a highly polished glass disk. The master is then strengthened and used to make a negative copy, or submaster, which is then used as the mold to make positive disks. These are given a coating of aluminum to make them reflective, and then finished with a protective coating of plastic.

Laser disks were first designed for use as video disks. However, during the 1970s they had to compete with video tape, which was rapidly improving in quality and could be used for recording as well as playback. As a result, video disks had only limited success. Recently, however, several companies have launched a new system known as Compact Disk Video (CDV). A CDV disk combines high quality digital sound with near perfect television pictures, and a CDV player can be played through both television and hi-fi equipment.

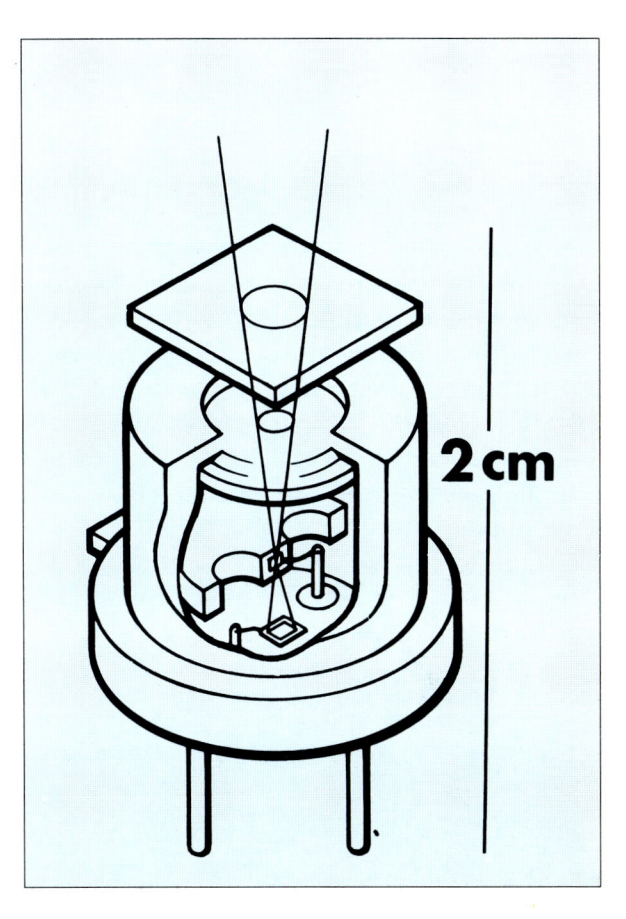

Using modern electronic materials it is possible to build very small lasers for use, in compact disk players.

The signal on a laser disk is recorded as a spiral of minute pits. The spirals are very close together; 1 μm, or micrometer is one millionth of a meter. Each pit and space corresponds to a single digital pulse. The reflected laser beam thus contains information about a series of digital pulses, and this can be converted into an electrical signal that corresponds to a sound signal.

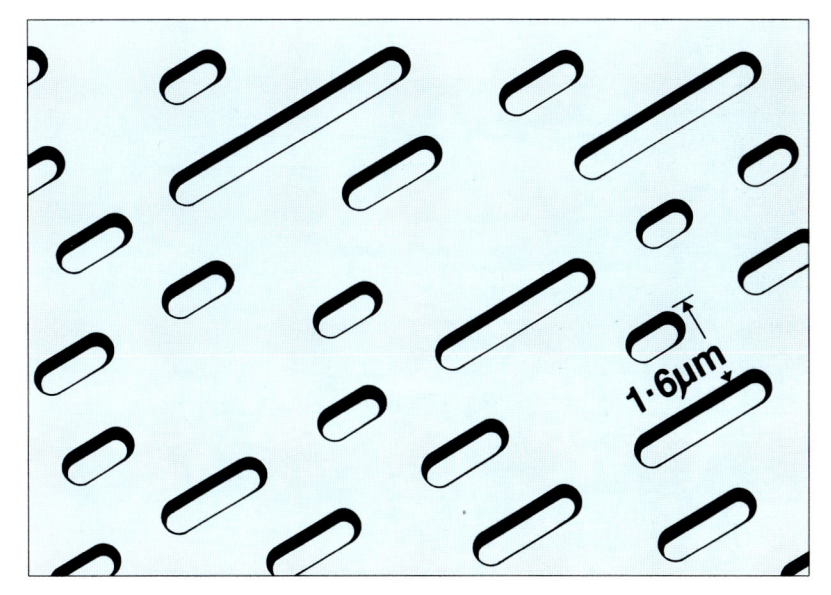

16 Audio systems

Audio systems have come a long way since the invention of the record player over 100 years ago. Technological developments have resulted in a steady improvement in the quality of broadcast and recorded sound. Modern audio systems are often referred to as hi-fi systems. This term was introduced in the 1960s, along with innovations like transistors, cassette tapes and stereophonic recording. The word hi-fi is short for "high fidelity" and reflects manufacturers' constant search for the most faithful sound reproduction that technology can create.

A modern audio system combines a number of different systems into one. At the heart of the system is the amplifier. This amplifies, or strengthens, the sound signals and generates a signal current strong enough to power the speakers. A wide range of different amplifiers are available. In the most sophisticated systems there may be a separate control amplifier, or pre-amplifier, and a power amplifier. Other systems use a single amplifier. The simplest systems have a receiver in which the amplifiers are combined with a radio tuner.

An audio system can be made up of a range of different hi-fi equipment.

The amplifier system is also used to control the mixture of bass (low) and treble (high) sounds. Also, there may be electronic filters that remove unwanted noises caused by interference or distortions in the recording equipment. Many systems have a graphic equalizer. This device controls sound quality by using special equalizer circuits. The manual controls of the graphic equalizer enable the user to alter the response of the system to different sound frequencies.

Most audio systems have a tuner. The tuner is basically a radio receiver, although most modern tuners have sophisticated circuitry for selecting stations and responding to weak signals. Other parts of an audio system are those used for playing prerecorded material, such as a turntable, a tape player and a compact disk player (see page 34). Modern high-precision turntables are designed to obtain the best possible results from standard records. Tape players have a variety of features, such as fast tape copying, tape search and separate heads for recording and playing. The Dolby system for reducing background noise is now standard on the best tape players.

The quality of sound reproduction depends a great deal on the quality of the speakers. Most systems have two speakers for producing stereo sound. Each speaker contains up to three units, which separately produce the low, medium and high frequency components of the sound.

Compact disk players are now available in a wide variety of forms. Here are shown two portable versions, one with built-in twin speakers (right) and one designed for use with a pair of headphones.

17 Television unlimited

Throughout the world television has become an almost essential part of people's lives. For millions of people, watching television is now the most popular form of recreation. In the United States in 1989, TV viewers watched an average of eight hours each day.

People watch television to be informed or entertained. At the same time, however, such audiences provide a huge potential market for those whose business it is to sell things. Advertising has become a key part of television, and except for a few public television stations, most television companies pay for their programs by selling advertising time. In some cases a whole program may be sponsored by just one organization. Advertisements have themselves started to become a form of entertainment; viewers tend to remember the products by amusing or dramatic advertisements, created by highly skilled film makers.

In many countries it is now possible to watch television at any time of the day or night. Viewers have an increasing choice of television channels to watch. In the past this choice was limited to a few television stations that transmitted signals along the ground, using booster relay stations where necessary. But for over twenty-five years it has been possible to transmit television signals around the world using communications satellites. However, there is a limit to the number of television stations that can operate in this way, since the available frequencies are rapidly being used up.

Modern television cameras are very light and portable. Here a camera operator is filming an event at the Los Angeles Olympic Games.

Satellite communications have made it possible for viewers all over the world to watch events as they happen and wherever they happen. Here a television camera joins the media as they record the historic visit to China of Queen Elizabeth II of England in 1988. Video recordings of the events were shown nightly on British television news.

One way of overcoming this problem is to use direct broadcast satellite television, which is now being used to widen people's choice of viewing still further. A direct broadcast satellite (DBS) is owned and operated by a satellite company, which may itself be made up of a group of smaller companies. The satellite company sells or leases channels to broadcasting companies, who buy their programs from program makers. Using cables (in some cases telephone links), the broadcasting companies transmit the programs to the transmission headquarters of the satellite company. From there, the program signal is "uplinked" (beamed up) to the satellite via the nearest suitable transmitter. The satellite amplifies and rebroadcasts a signal that is sufficiently powerful to be picked up by a relatively small receiver on or near the house of each viewer. Some channels are free; others can only be seen by viewers who have paid a subscription fee to the satellite company.

Today's DBS systems are not ideal. To receive all the channels that are available, people are having to buy several different antennas and a number of different receivers and decoders. At the same time the technology behind most of the equipment is progressing too fast for manufacturers to keep up. Equipment is becoming out of date almost before it leaves the factory. All of this is very frustrating for manufacturers and confusing and expensive for the viewer.

An alternative to DBS is cable television. A cable television company receives signals from several different sources, including satellites and ground-based broadcasters. The programs are then sent to subscribers via an optical-fiber cable buried underground. The viewer pays the cable operator a monthly fee without having to buy any special equipment. Cable television is widespread in the United States and could in time become more popular in Britain and Europe.

A prototype videophone, a telephone combined with a television set.

TELEVISION UNLIMITED

Television is a powerful publicity medium. Here cameras record pictures of President George Bush and his cabinet.

But is the increase in the amount of available television a good idea? Many people argue that it is not. They maintain that the quality of programs declines; that the more expensive dramas and documentaries disappear, together with responsible news and current affairs programs. This leaves a television choice of soap operas, cheap game shows, second rate movies and sensation-seeking news.

Others argue that television already has too much influence on our lives. It is said that, as a result of television, we read less and spend less time talking to one another. Many people believe that the presence of television cameras at certain events, such as political demonstrations, actually influences what happens because of the potential publicity. There is little doubt that television can influence what we think and do – as politicians and advertisers are very well aware. And as a result people are becoming more and more concerned about the content of some of the programs that are broadcast, particularly those that contain scenes showing violence.

On the other hand, there is probably a limit to how people's opinions and ideas can be influenced by what they see on television. Most people are able to distinguish between reality and fiction, although it is clear that both program makers and parents must take responsibility for what young children see on television. Many television programs open up a world that most of us would never otherwise have a chance to see. The debate about television and its effect on our lives will go on for a long time to come.

41

⬥ 18 Multichannel communications

Predicting the future of television, video and audio systems is not easy. Ideas that look promising now may fail to materialize for any one of a number of reasons, including high cost, more popular alternatives or simply the fact that the technology fails to work properly. The black-and-white LCD wristwatch televisions of the early 1980s proved to be little more than gimmicks. Perhaps even today's miniature color televisions may find only a limited market. Another idea put forward recently involves using laser beams to play standard records without wearing them out. But this looks like being both unnecessary and impractical.

Today, Direct broadcast satellite (DBS) television can be received by individual homes, via a small dish antenna set up on or near the house.

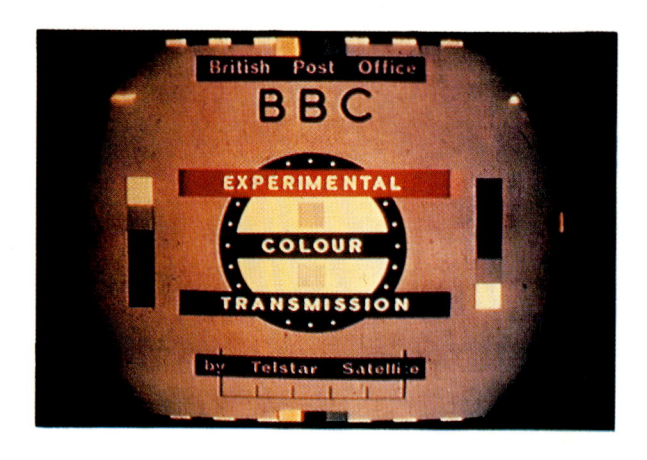

The first intercontinental television pictures were transmitted via the satellite *Telstar* in 1962.

42

A cable television studio. The studio passes programs to subscribers by cable.

On the other hand, some of today's ideas may well prove successful. Scientists are working on ways of producing compact disks that can be used for recording as well as playing music. Such disks could then compete with DAT and, ultimately perhaps, CDV recording systems might replace videotape recording systems.

The idea of flat television sets that occupy less space than today's bulky cathode ray devices appeals to many people, and it now seems likely that LCD technology will eventually provide us with full-sized flat television sets. At the same time the parts of audio systems may also become smaller, making them easier to position in a room and, if desired, easier to conceal. As a result television and audio systems need not dominate people's living areas as they do now.

For some time scientists have also been working on ways of generating three-dimensional television. Ideas range from using special grooved screens to generating moving holograms. However, it remains questionable as to whether people will really want three-dimensional images in their living rooms.

The television and audio systems of the future will probably not remain the isolated systems they are today. Modern electronic digital control has made it possible to link together systems that were formerly incompatible. We are gradually progressing toward the time when all the systems of the home will form a single unit. At the heart of the unit will be a computer, and the whole system will be operated from a central console, equipped with a keyboard, monitor, telephone, printer facsimile machine and other controls. The computer will have two-way links to all telephones, television sets, video recorders and audio systems, and will also control such things as the central heating system, the security system, and fire and fault detecting systems. Television, video and sound will then just be a part of a single, multichannel communications and control system.

Glossary

Amplitude. The distance between the center and top, or center and bottom of a wave.
AM. Amplitude modulation. A method of transmitting a radio signal in which a sound or video signal is made to vary the amplitude of a carrier wave.
Amplifier. A device using transistors to increase the power of a radio signal.
Analogue. Something that is similar to or corresponds to something else. An analogue electrical signal is one that varies in a wave-like way that corresponds exactly to the sound signal that produced it.
Antenna. A metal device used for converting radio signals into electric currents. Tuning circuits in the receiver select the current of the desired radio frequency.
Bar code. A row of black and white stripes that represent numbers in computer code. The stripes can be scanned by a laser beam and the reflected beam is converted into a signal that can be interpreted by the computer.
Betamax. A type of video cassette recorder, now largely out of fashion.
Camcorder. A combined camera and video recorder.
Cathode ray tube (CRT). A device for generating a beam of electrons (cathode rays) and focusing it onto a fluorescent screen.
CD. Compact disk. A form of sound recording in which the audio signal is recorded in the form of pits on a reflective surface. During playback, the pits are scanned by a laser beam.
CDV. Compact disk video. A laser-read disk, similar to a compact disk, on which both sound and video signals are recorded.
Communications satellite. A satellite used for receiving, amplifying and retransmitting radio signals.
Conductor. A material through which an electric current can flow easily.

DAT. Digital audio tape. Very high quality tape used for sound recording.
DBS. Direct Broadcasting Satellite. A form of communications satellite used for transmitting television signals directly to subscribers on the ground.
Digital. Concerned with numbers or digits. A digital signal is one that has been coded into a sequence of numbers, which can then be converted into computer code.
Diode. A thermionic valve or semiconductor device that converts alternating current into direct current.
Electrode. A conductor by which electric current enters or leaves a device such as a thermionic valve, cathode ray tube or electric cell ("battery").
Electromagnetic radiation. Radiation composed of waves of energy generated by vibrating electrical and magnetic fields.
Electron. One of the tiny negatively charged particles that form part of every atom.
ENG. Electronic news gathering. The use of video cameras and video recording to record and transmit news and other events.
Fluorescent. Term often used to describe a material that absorbs energy, in the form of moving electrons or electromagnetic radiation, and gives out energy in the form of light.
FM. Frequency modulation. A method of transmitting a radio signal in which a sound or video signal is made to vary the frequency of a carrier wave.
Frequency. The number of cycles, or oscillations (to-and-fro movements) that occur in a given time.
FST. Flatter squarer tube. A television picture tube with a relatively flat screen for reducing picture distortion.
Infrared. Invisible electromagnetic radiation of slightly longer wavelength (and therefore lower frequency) than red light.

LCD. Liquid crystal display. A screen that uses liquid crystals, whose light-transmitting properties alter as they are subjected to changing electrical voltages.
Magnetic field. A field of force that exists near a wire carrying an electric current or near a permanent magnet.
Microphone. A device used to convert sound waves into a varying electric current.
Monitor. A screen linked by a cable directly to a computer or television camera.
Phosphor. A substance capable of absorbing energy and releasing it in the form of light.
Semiconductor. A material that can, under different circumstances, behave as a conductor of electricity or as a non-conductor (insulator).
Stereophonic sound. Sound recorded using two separate channels, which are replayed through two speakers. To a listener positioned correctly in relation to the speakers, the sound appears to have a "three-dimensional" quality.
Teletext. Computer information broadcast by a television company as part of the video signal.
Transistor. An electronic device, made up of three layers of a semiconductor material, that performs the same functions as a triode valve, but is more reliable and consumes less power.
Tuned circuit. An electrical circuit that resonates to a particular frequency. In its simplest form it consists of a coil and a capacitor (a device that stores electric charge).
Tuner. A device for tuning into different radio stations.
UHF. Ultra high frequency radio waves.
VCR. Video cassette recorder. A machine that records and plays video signals by means of tapes contained in cassettes.
VHF. Very high frequency radio waves.
VHS. Video Home System. Currently the most popular format used for video cassette recording.
VHS-C. A type of camcorder in which a short length of standard VHS tape is contained in a small cassette. The tape can be replayed in a video recorder using an adaptor.
Videotex. Computer information, stored on central computers, that can be displayed on a television screen. *See* **teletext** and **viewdata**.
Viewdata. Computer information compiled by a viewdata service and transmitted to subscribers via the telephone.

Further reading

Alden R. Carter, *Radio: From Marconi to the Space Age* (Franklin Watts, 1987)
Wayne J. LeBlanc & Alden R. Carter, *Modern Electronics* (Franklin Watts, 1986)
Robin McKie, *Technology: Science at Work* (Franklin Watts, 1984)

James L. Schefter, *Telecommunications Careers* (Franklin Watts, 1988)
Glenn Alan Cheney, *Television in American Society* (Franklin Watts, 1983)
Mat Irvine, *TV and Video* (Franklin Watts, 1984)
Carolyn Cooper *VCRs* (Franklin Watts, 1987)

Picture Acknowledgments

The publishers would like to thank the following for allowing their photographs to be reproduced in this book:
Allsport 20, 38; Aquarius 18; Topham 5, 7 (above), 19 (both), 23 (both), 26, 28, 29, 30, 31, 34, 36, 37 (above), 39, 40, 41; TRH (The Research House) 4, 6, 10, 12 (both), 16, 22, 24, 35 (both), 37 (below), 42 (both), 43; Visnews 21; ZEFA 7 (below).
Artwork by the Hayward Art Group, apart from illustrations on pages 8, 32, and 33 by Nick Hawken.

Index

advertising 38
alternating current 11
AM (amplitude modulation) 8, 9
amplifier 36, 37
analogue recording 24, 35
antenna 9, 14, 21, 22, 40
Armstrong, Major Edwin 7
Atlantic Ocean 6
atmosphere 9
audio 5, 29, 36, 37
audio casettes 29

Baird, John Logie 7
banks 23, 30
bar codes 27
Berliner, Emile 32
Betamax 26
binary code 25
Bologna 6
Branly, Edouard 6
Braun, Ferdinand 11

cable television 5, 40
calculator 10, 16
camcorder 29, 30
camera 12, 13, 18, 21
carrier wave 8
cathode ray tube 7, 11, 12, 14, 16, 22
CDV (compact disc video) 35, 43
closed circuit television 30
coherer 6
communications 4, 5
communications satellites 38
compact disk 32, 34, 37, 43
computers 5, 10, 11, 22, 23, 25, 31, 35, 43
conductor 10
control room 18, 19

DAT (digital audio tape) 25, 43
DBS (direct broadcast satellite television) 39, 40
decoder 23, 40
de Forest, Lee 11
diamond 32
digital recording 25, 35
digital watch 10, 16
diode valve 11
direct current 11
director 18, 19
Dolby system 37

Edison, Thomas 10, 11, 32
electromagnetic head 24
electromagnetic radiation 8
electron beam 12, 13, 14, 16
electron gun 14
electronics 10, 11, 27
electrons 10, 15
ENG (electronic news gathering) 21
entertainment 4, 38
entertainment technology 18, 19
erase head 24

Fastext 23
Fessenden, Reginald 6
Fleming, John 10, 11
fluorescent screen 14, 23
FM (frequency modulation) 8
frequency 8

gamma rays 8
generators 21
graphic equalizer 37
grid 15

Hertz, Heinrich 6
hertz 8
hi-fi equipment 10, 35, 36

holograms 43

iconoscope 13
information 22, 23
infrared 17, 26
ionosphere 9
iron oxide 24

JVC 26

lasers 27, 34, 35, 42
LCD (liquid crystal display) 16, 42, 43
lens 12, 13
light rays 8
long waves 8
LP (long playing) disc 32

magnetic tape 22, 24, 25, 29
Marconi, Guglielmo 6
master disc 32, 35
master tape 32
medium waves 8
microchips 17, 23, 25
microphone 6, 8, 19, 32
microwaves 8
mixing 19
modulation 8
monitors 19, 22, 43
Morse code 6
movies 28
multichannel communications 42, 43

navigation 4
news programs 18, 20, 41
Nipkow disk 7

optical-fiber cable 40
orthicon 13
outside broadcast 20, 21

46

Philips 26
phosphors 15, 16
picture cells 12, 16
picture receivers 14, 15, 16, 17
pixels 12, 13
police 30
primary colors 13

radar 4
radio 4, 5
radio waves 6, 7, 8, 9
recording sound 32
record/play head 24, 29
remote control 17, 26, 31
reproduction 34
robbery 30
robots 30, 31

sapphire 32
satellite television 5, 7, 39, 40
scanning 13
screen 13, 16
security 30, 43
sending pictures 12
sensor 17

shadow mask 15
Shockley, William 11
short waves 8
silicon chips 10, 11
Sony 26
sound engineers 13
space exploration 31
stereophonic sound 5, 7, 32
studio 13, 18, 19, 20, 32
stylus 33, 34

tape recorder 24
technical director 19
telegraph 4
telephone 4, 6, 23, 43
teleprompt 18
teletext 23
television camera 12, 13, 20
television receiver 14, 15, 16, 17
thermionic valve 10
Thomson, J. J. 10
transistor 7, 10, 11
transmission 6, 16, 21
transmitter 13, 39
triode valve 11

tuned circuit 6
tuner 37

UHF (ultra high frequency) 8, 9

valves 11
VCR (video cassette recorder) 26, 27
VDU (visual display unit) 22
VHF (very high frequency) 8, 9
VHS (video home system) 26, 29
video camera 20, 29
video conferencing 30
video recording 13, 26, 27, 28, 30
videotape 18
Videotex 23
vidicon 13, 29
viewdata 23

wavelength 8, 9
wireless 4

X-rays 8

Zworykin, Vladimir 7, 13

47

PROPERTY OF
LEWIS ELEMENTARY SCHOOL
1431 N. LEAMINGTON AVE.
CHICAGO IL 60651